Quizzes, QUOTES, QUESTIONS, and MORE!

BY ARIE KAPLAN

ILLUSTRATED BY Risa Rodil

Grosset & Dunlap

GROSSET & DUNLAP
An imprint of Penguin Random House LLC, New York

First published in the United States of America by Grosset & Dunlap,
an imprint of Penguin Random House LLC, New York, 2024

Photo credits: used throughout: (speech bubbles with question marks)
Oleksandr Melnyk/iStock/Getty Images

Visit us online at penguinrandomhouse.com.

Manufactured in Canada

ISBN 9780593888995 10 9 8 7 6 5 4 3 2 1 FRI

Design by Kimberley Sampson

TABLE OF CONTENTS

BIRTH OF THE BEAST

Who *Is* Mr. Beast?

Jimmy Donaldson, better known as MrBeast, is one of the biggest YouTubers in the world. He makes fun, exciting, kinetic videos chock-full of stunts, pranks, increasingly ambitious production values, and cash giveaways. MrBeast's videos are popular because he gives his viewers access to things and places they would never see in their everyday lives. In one video, he's chased by an FBI agent. In another, he visits luxury hotel rooms around the world.

MrBeast is also very serious about helping people who are less fortunate through his many charitable endeavors.

But who *is* MrBeast, really? How did he get started as a YouTuber, a prankster, and an entrepreneur? You might not believe this, but before he was racking up millions of views and devising larger-than-life stunts, MrBeast was just an ordinary kid with a laptop and a dream.

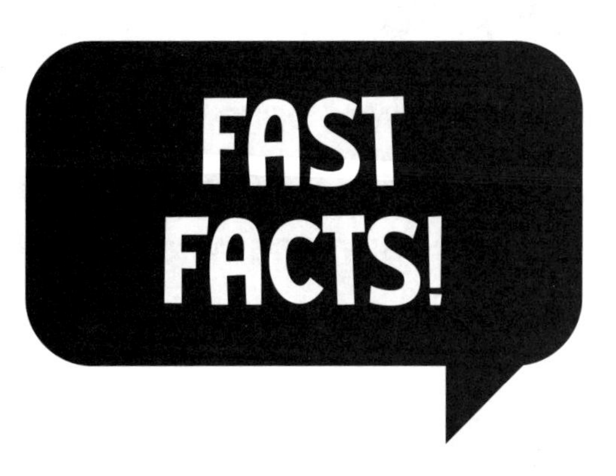

At the end of 2021, according to Guinness World Records, MrBeast held the record for the YouTube contributor with the highest earnings.

That year, MrBeast earned an annual income of $54 million.

Born in Wichita

Jimmy Donaldson was born on May 7, 1998, in Wichita, Kansas. Both of his parents were active duty military, and when Jimmy was growing up, the family moved around quite a bit because of this.

In 2007, when Jimmy was eight years old, his parents divorced. After that he lived with his mother, Sue.

When Jimmy was a child, he was somewhat quiet. According to his childhood friends, one way to get him to open up was to talk to him about his two main interests, gaming and YouTube.

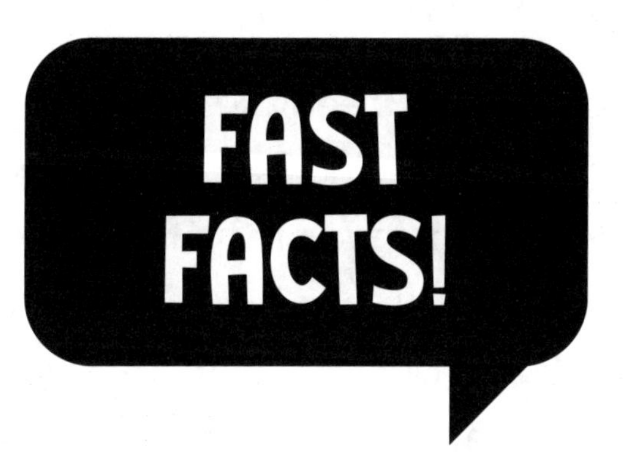

Jimmy's mother, Sue, is a former prison warden.

Jimmy grew up playing baseball with Chandler Hallow, who would eventually costar in many MrBeast videos.

MrBeast6000

Jimmy grew up in Greenville, North Carolina. Because he was such a big fan of YouTube, in February 2012, at the age of thirteen, Jimmy uploaded his first YouTube video under the username MrBeast6000. It didn't get very many views. He wondered what he would have to do to be successful as a YouTuber.

Shortly afterward, Jimmy began attending high school at a private school called Greenville Christian Academy. During this period, he also loved playing baseball with his friends. However, during his sophomore year of high school, Jimmy was diagnosed with Crohn's disease, which is an autoimmune disorder. Because of that, he had to stop playing sports.

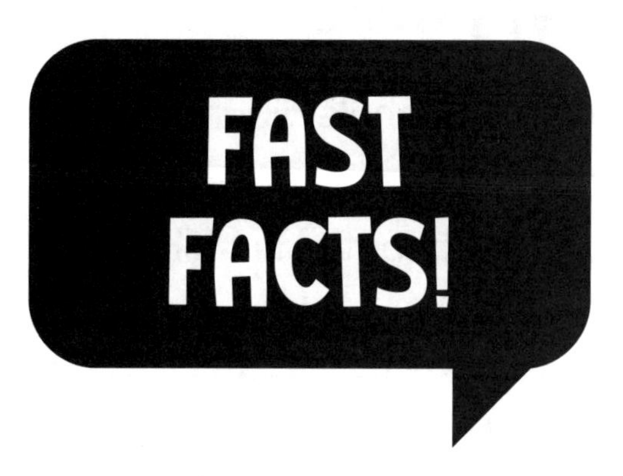

These days, Jimmy manages his Crohn's with medication.

He also works with a private chef because sticking to a very specific diet can help him manage flare-ups.

Trying Everything

After Jimmy's Crohn's diagnosis, he continued posting videos under the MrBeast6000 username. He still wasn't getting very many views. How could he master the YouTube algorithm? What kind of content would attract more eyeballs?

MrBeast decided to try a little bit of everything on his channel, to see if anything would get him a big audience. Some of his videos showed him playing *Call of Duty* and *Minecraft*. In other videos he tried to guess how much money his fellow YouTubers made. He was experimenting with so many gimmicks, so many *types* of videos. But nothing was clicking.

Would that change? Would Jimmy ever figure out how to be a successful online personality?

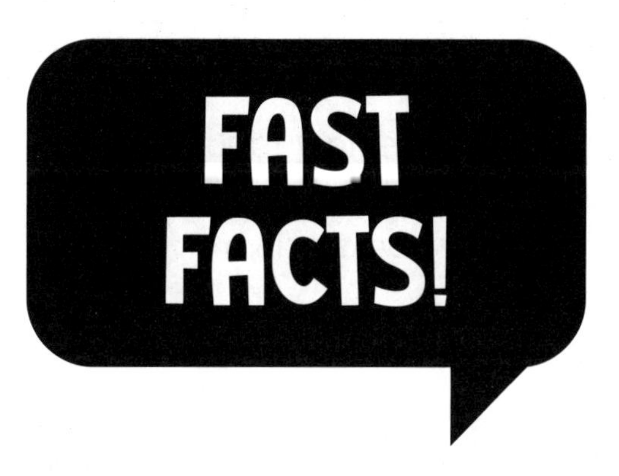

Jimmy started uploading videos as MrBeast6000 because the username "Mr. Beast" was taken.

In these early videos, Jimmy didn't show his face on camera that often.

Did You Know That . . .

1. Jimmy respects how tough his mother, Sue, had to be in order to be a prison warden.

2. Today Sue works as the chief compliance officer for Jimmy's company, MrBeast LLC.

3. Jimmy's middle school teacher has described him as a conscientious student.

4. Jimmy has a stepfather whose name is Tracy.

5. Tracy is an operations manager of MrBeast LLC.

6 Jimmy and his parents lived in three different locations in the American South by the time he was seven years old.

7 But for the most part he was raised in Greenville, North Carolina.

8 And he considers Greenville his hometown.

9 Jimmy still lives in Greenville today!

10 Greenville is a small, beautiful city where you can find many office parks and strip malls, among other things.

Early Days

"When I was a young teenager, I was getting no views, had no money, had no equipment . . . I was using my brother's old laptop. So my first couple hundred videos, I didn't have a microphone."

—Jimmy on his early days of recording

If you were a YouTuber who was just starting out, where would you get your equipment? Would you raise money for it? Would you borrow someone else's stuff? Would you ask your friends or family members for help? Write about it on the lines below.

Passion Project

Jimmy is passionate about being a YouTuber. What are you passionate about? What do you like to do more than anything else in the world? Do you like to sing? Do you like to dance? Do you like to write short stories or poems? Do you like to draw or sculpt? Do you like to play football? Do you like to solve math problems? Whatever you're passionate about, write about it on the lines below.

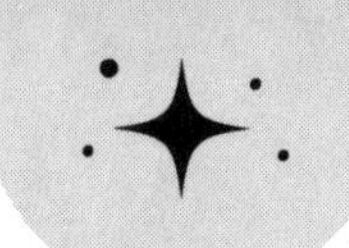

Quick Quiz: Just Starting Out

1) Jimmy's zodiac sign is ____.

a. Pizza
b. Cheeseburger
c. Taurus
d. Fries

2) Even before Jimmy started using the "MrBeast6000" handle at age 13, he'd already begun uploading YouTube videos when he was ____.

a. 11 years old
b. 64 years old
c. 87 years old
d. 120 years old

3) Jimmy's mother Sue has a huge warehouse where she saves every ____ from his childhood.

a. Snickers bar
b. Three Musketeers bar
c. Milky Way bar
d. memento

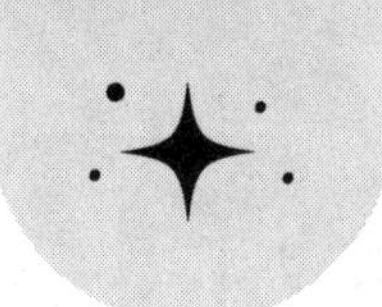

4) **One of Jimmy's heroes is Apple founder ____.**

a. Abraham Lincoln
b. Steve Jobs
c. George Washington
d. Thomas Jefferson

5) **MrBeast likes reading books about successful people like basketball superstar Michael ____.**

a. Cupcake
b. Pie
c. Jordan
d. Cannoli

Check your answers on page 78!

FINDING HIS VOICE

Cracking the Code

In 2015 and 2016, things began to change for Jimmy. That's when he made a series of "worst intro" videos, making fun of the introductions other content creators were making for *their* videos. Jimmy started to become known for these "worst intro" videos. He hit 30,000 subscribers by the summer of 2016.

He was finally figuring out what worked. And he was finding his voice: somewhat snarky and sarcastic, but also likable and funny. Most importantly, he was beginning to be himself on camera and be more relaxed. He was introducing audiences to the *real* Jimmy Donaldson.

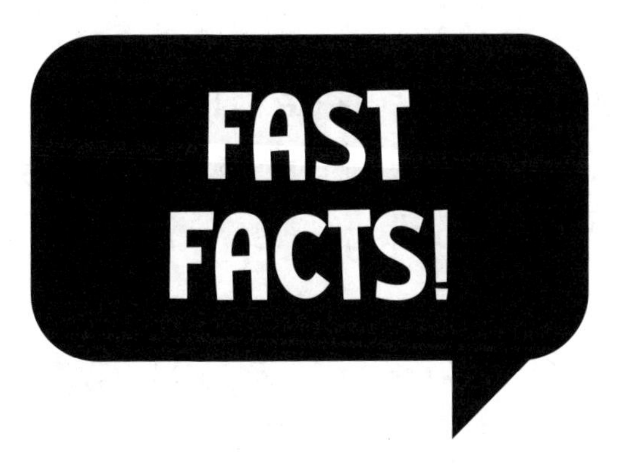

For years, Jimmy's mother didn't know about his YouTube videos.

Then she saw them referenced in his high school yearbook.

College?

Jimmy enrolled in college in late 2016. However, he dropped out only two weeks into his first semester. College just wasn't for him. Besides, his career as a YouTuber was really starting to take off. He wanted to stay focused on it.

At the time, he was eighteen years old and living with his mom in his childhood home. Jimmy and his mother were always close. However, she was upset that he dropped out of college, and kicked him out of the house. But she was happy to see that he was having success as a YouTuber.

FAST FACTS!

Even though Jimmy just briefly attended college in 2016, he taught a class at Harvard Business School in 2023.

That same year, with characteristic good humor, he posted on Instagram about the irony of that situation.

Countdown to Success

Shortly after he dropped out of college, Jimmy moved into a duplex with his childhood friend.

But although his videos were getting a decent number of views, he knew that he had the potential to do better. Then, in January of 2017, Jimmy uploaded a video that was unlike any he'd ever made before. It showed him looking straight at the camera and counting from zero to one hundred thousand in one sitting. It quickly became his first viral video.

(And because many journalists wrote articles about the "counting to 100,000" video—and they referred to Jimmy as MrBeast in articles *about* the video—*we're* going to refer to him as MrBeast from now on as well.)

The video had taken MrBeast forty hours to make. It taught him that people really liked it when he did silly stunts. And he decided that maybe he should do that more often.

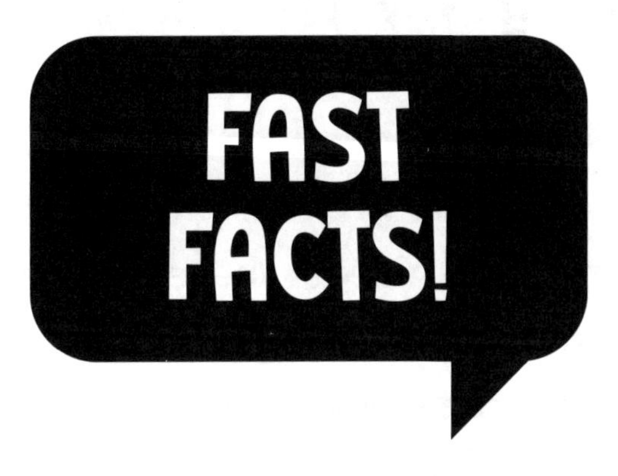

The "I Counted To 100,000!" video amassed more than 6.5 million views when it was first uploaded in January 2017.

That video is *long*. It lasts for nearly twenty-four hours.

Reaching New Heights

With that first viral video in January 2017, MrBeast finally found out what kind of content the YouTube algorithm craved: extreme stunts. Soon he made more videos where he did stunts. In one of them, he spun a fidget spinner for twenty-four hours straight. In another, he watched the "It's Everyday Bro" music video by Jake Paul over and over again on a continuous loop for ten hours straight.

Then, it happened. In November of 2017, MrBeast hit one million subscribers. He was making his dreams come true and had become very successful in the process. Ever since then, his videos have only become more ambitious and inventive, and in many cases, more epic!

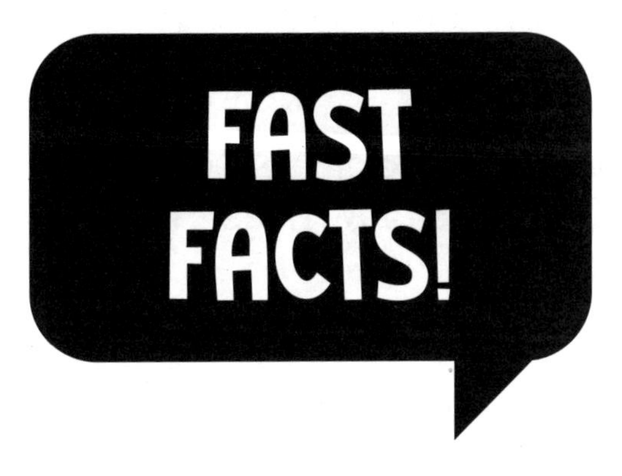

Jake Paul is a YouTuber, a social media influencer, and a boxer.

In 2022, Jake said that he could never compete with MrBeast's YouTube empire.

Did You Know That . . .

1. Around the time of the "I Counted To 100,000!" video, MrBeast reached 750,000 subscribers.

2. He also got his first brand deal during that time.

3. But he didn't spend the money he made from the deal.

4. Instead, he invested it in a video where he gave an unhoused person a check for $10,000.

5. This wasn't MrBeast's first video where he gave away free money.

6 And it wouldn't be his last.

7 The original cut of the "I Counted To 100,000!" video was forty hours long.

8 But due to the constraints of the video editing software MrBeast was using, he had to make it slightly less than twenty-four hours long.

9 For that reason, some parts of the video are sped up.

10 As of early 2024, it has over 30 million views.

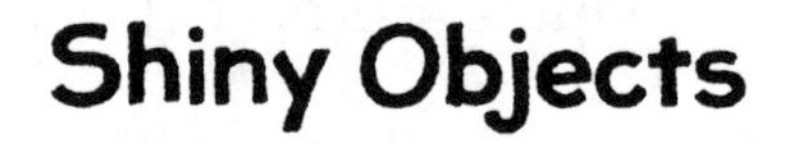

Shiny Objects

"I don't want to live my life chasing the next shiny object to the next shiny object."

—MrBeast on not placing very much importance on material possessions

Was there a time in your life when you were obsessed with a material possession? What was it? A piece of clothing you saw in a store window? Jewelry? A video game? How did you come to realize that it wasn't that important? Write about it on the lines below.

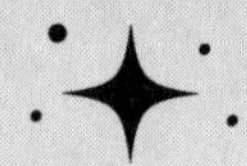

Talk to Future You!

"I hope you have at least 100k subs," MrBeast said in a 2015 video for his future self. (Two years later he would have a million subscribers.)

If *you* made a video for *your* future self, what would it be about? What would you tell yourself? Write about it on the lines below.

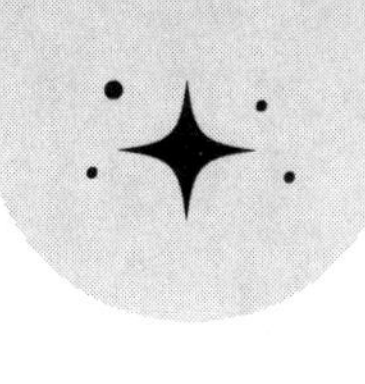

Quiz: Leveling Up

1) In one early stunt video, a friend wraps MrBeast in one hundred layers of ____.

a. eels
b. toilet paper
c. snakes
d. dragons

2) In one 2018 video MrBeast and several other people all walked into random stores wearing inflatable ____ costumes.

a. dinosaur
b. beagle
c. poodle
d. bulldog

3) The college MrBeast briefly attended was called ____.

a. Monsters University
b. Starfleet Academy
c. East Carolina University
d. Time Lord Academy

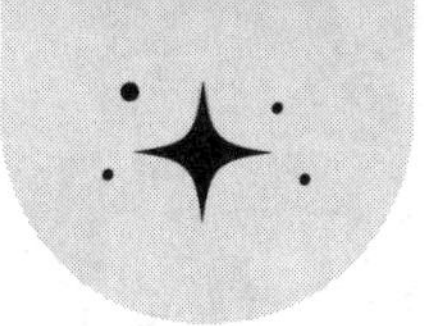

4) Stunts where a YouTuber buys as much of a specific item as possible or spends an exorbitant amount of money on a specific product are commonly known as ____ videos.

a. nice

b. great

c. wonderful

d. junklord

5) MrBeast is a fan of the strategy game Settlers of ____.

a. Swiss cheese

b. Catan

c. American cheese

d. Brie

Check your answers on page 78!

THE MOST MEMORABLE VIDEOS

Slimy Houses and Cereal Bowls

Over the past few years, MrBeast has made so many amazing videos. In one of them, he filled his brother CJ's house with barrels and barrels of slime! And not just any slime, a kind of rapidly expanding foamy goop that shoots through windows and doors like a battering ram. By filling CJ's house with slime, MrBeast was playing an outrageous prank on his brother. But he made it up to CJ by buying him a new house.

In another video, MrBeast and his crew (Kris, Chandler, and Jake) filled a humongous bowl with expired cereal and powdered milk. Then they picked random people to sit inside the gigantic cereal bowl. The last person to leave the bowl won $10,000.

And sure, these videos are both very memorable. That goes without saying. But what are the *most* memorable ones MrBeast has ever made?

Keep reading, and you'll find out!

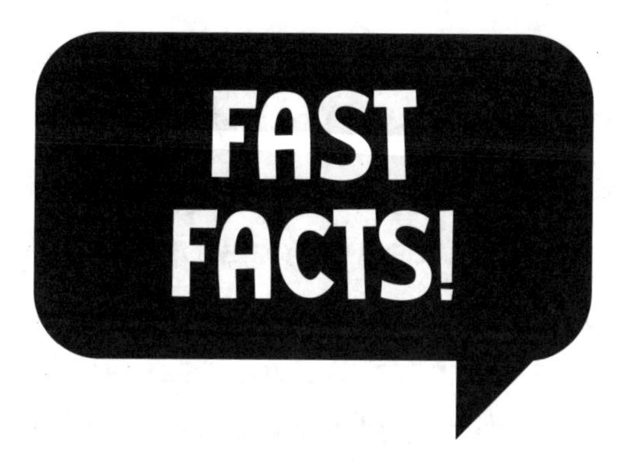

The winner of the "World's Largest Cereal Bowl" challenge was named Scotch.

Scotch managed to stay inside the bowl for twenty hours straight.

Only One Minute!

In the video titled "I Gave People $1,000,000 But ONLY 1 Minute To Spend It!," MrBeast went to various locations—grocery stores, jewelry stores, electronics stores, a video game store—and gave people the opportunity to spend a million bucks in sixty seconds. Aside from the very severe time limit, the other rule for this challenge was that MrBeast would pay for whatever the contestants spent.

Nobody was able to spend the entire million dollars in just one minute, which is—to be fair—a *very* short amount of time in which to do *anything*.

MrBeast also switched up the very premise of the game for some segments of the video. For instance, he lined up cars on a football field and told a contestant, Mitch, to run toward the car he most wanted as the countdown began. If Mitch could drive the car back across the finish line before a minute had passed, he got to keep the car. Mitch came away from the experience with a brand-new Tesla.

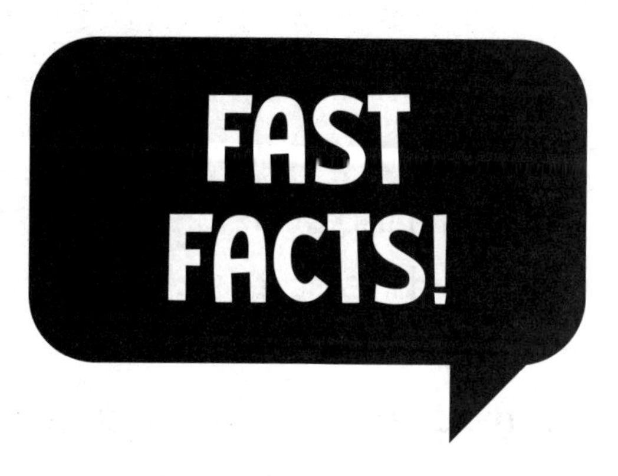

The type of Tesla that Mitch chose is worth roughly $40,000.

In the jewelry store portion of the video, the contestant, Ashley, bought $60,990 worth of bling.

Underwater

In MrBeast's "Spending 24 Hours Straight Under Water Challenge" video, he was submerged in a swimming pool. His head was inside a chamber. And inside the chamber there was an oxygen pump.

As MrBeast made observations about what it was like underwater, his friends would occasionally swim into his chamber to chat with him and keep him company. And when his phone died after six hours, they bought him a new one. However, after about twelve hours in the underwater chamber, MrBeast began to feel seasick, and he was taken back up to the surface by his friends.

But it needs to be said that twelve hours underwater is quite an achievement!

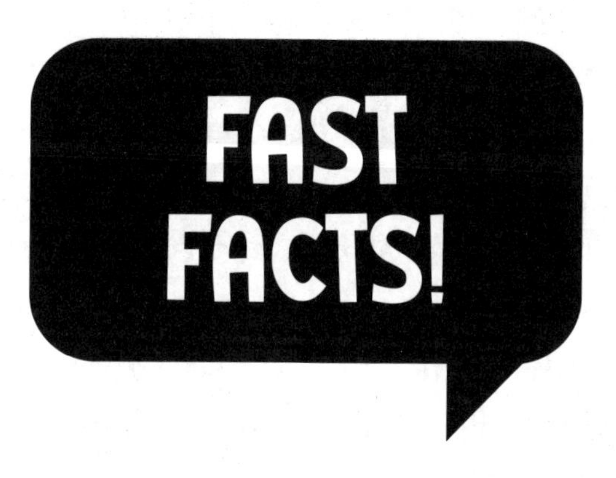

At the end of the video, MrBeast made it a point to tell his viewers *not* to try to duplicate his underwater stunt.

As a role model for young people, MrBeast knows that he has a responsibility to model good behavior.

Squid Game IRL

The Korean television series *Squid Game* was about a mysterious competition where winners get a big cash prize. But if you lose the game, you lose your life.

The competition series depicted in *Squid Game* is thankfully fictional. All of the "contestants" were actors. But shortly after the September 2021 debut of *Squid Game* in the United States (via Netflix), MrBeast had a thought: What if it was real? And what if the consequences of losing were non-lethal? If you lose the game, nothing bad happens to you except that . . . well, you lose the game. With this in mind, in November 2021 MrBeast produced and uploaded a video titled "$456,000 Squid Game In Real Life!"

The version of Squid Game in MrBeast's video involved 456 actual contestants playing games much like the one on the Netflix series, but without the violence. However, the winner would still get a cash prize. It cost a whopping $3.5 million for MrBeast to create the video, in which contestants made their way around a large space decorated to look like the various sets from the show. Building the sets was hard work, but it was worth it. Just a few weeks after the video was first uploaded, it boasted 158.2 million views!

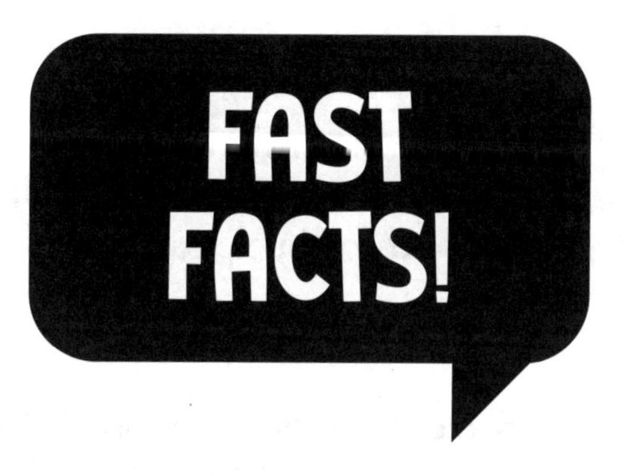

Every single contestant in MrBeast's "Squid Game" video had a device strapped to them.

And whenever a contestant was eliminated, the device popped (harmlessly).

Did You Know That . . .

1. Karl Jacobs, who is in many of MrBeast's videos, started out as an editor for the videos before being bumped up to on camera talent.

2. In the video game store portion of the "I Gave People $1,000,000" video, the contestant was named Sean.

3. Sean is Karl Jacobs's brother.

4. For a few moments in the "Spending 24 Hours Straight Under Water" video, MrBeast eats sushi.

5. Which is appropriate since sushi is seafood.

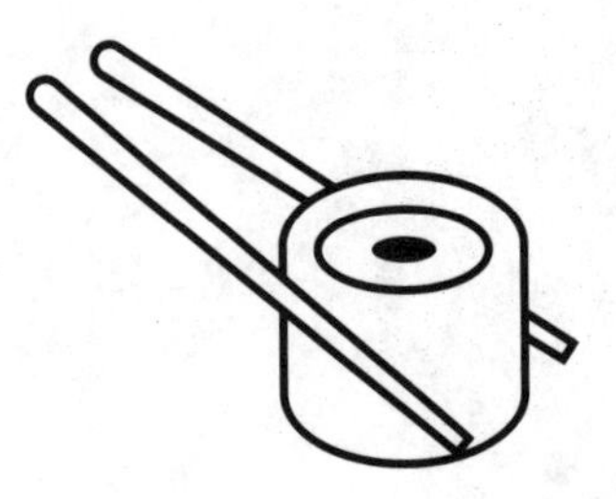

6 When the "$456,000 Squid Game In Real Life!" debuted, it was MrBeast's most-watched video ever.

7 MrBeast's "Squid Game" video is just under twenty-six minutes long.

8 And since that video cost $3.5 million to make, that means that *each minute* cost $134,600.

9 One of the contestants in MrBeast's "Squid Game" video was a fellow YouTuber who goes by the handle NightFoxx.

10 In November of 2021, *People Magazine* published photos of MrBeast and his team building and painting the sets for the "Squid Game" video.

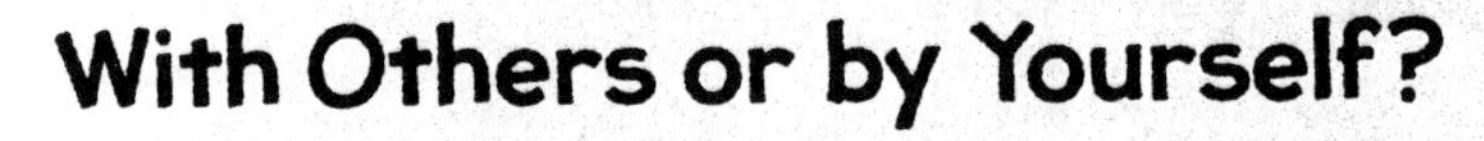

With Others or by Yourself?

"I've literally worked with over a thousand people."

—MrBeast on working with many friends, colleagues, and employees over the past several years. He has described himself as an introvert. However, as the owner of a thriving business, he collaborates with a large number of people on a regular basis.

Do *you* like working with other people? Or are you happier working alone? Write your answer on the lines below.

Last to Leave!

Aside from extreme stunts, one of MrBeast's most popular types of videos are the "last person to leave" challenges, where the last person to leave a certain area gets a cash prize. Sometimes, the challenge involves being the last person with their hand touching an object, like an expensive sports car. The last person touching the car *wins that car*!

If you could come up with a "last person to leave" challenge, what would it be about? Would it involve being the last person to leave an area? The last person to touch something expensive? What would that expensive thing be? A castle? A car? An airplane? A robot? A rare, hard-to-find toy? A rare, hard-to-find chocolate bar? Write about your "last to leave" challenge on the lines below!

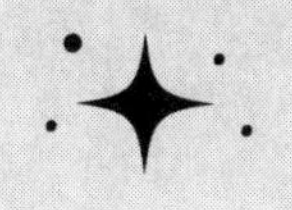

Quiz: How Well Do You Know the Beast Gang?

In MrBeast's videos, he's often joined by a group of friends. They're his costars, really, and many of them have their own YouTube followings. Two of them, Chandler Hallow and Karl Jacobs, have already been name-checked in this book. They're all mentioned in the quiz below, and so are a couple of MrBeast's *other* costars, who are sometimes known as the Beast Gang. And now it's time to answer the question: How well do you know the Beast Gang?

1) Karl Jacobs is from upstate ____.

a. Mars
b. Vulcan
c. Saturn
d. New York

2) In addition to his work as a part of MrBeast's crew, Tareq Salameh is a ____.

a. stand-up comedian
b. stormtrooper

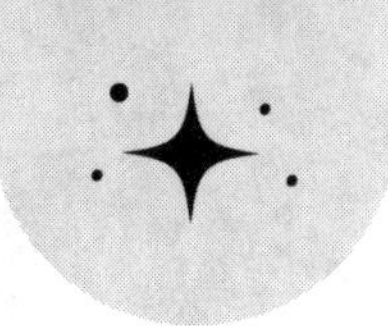

c. Jedi Knight
d. Sith Lord

3) Chandler Hallow made his on-camera debut in a MrBeast video in which year?

a. 1859
b. 1897
c. 2017
d. 1864

4) Before Chandler Hallow was a member of MrBeast's crew, he worked for the famed YouTuber as his ____.

a. sorcerer
b. janitor
c. wizard
d. warlock

5) Where is Nolan Hansen from?

a. Gotham City
b. Metropolis
c. The mythical island of Themyscira
d. Nebraska

Check your answers on page 78!

CHARITY AND KINDNESS

Giveaway

Aside from his outrageous stunts MrBeast is known for his random acts of kindness. In February 2021, MrBeast uploaded a video called "I Donated $300,000 To People In Need." He only gave the money to those who really needed it. For instance, he gave $30,000 to a restaurant owner who was going out of business. Then he paid $15,000 to the property manager of an apartment complex to cover all of her tenants' rent. Many of the people he gave money to had been affected by the COVID-19 pandemic, and they were very thankful for his help.

But if MrBeast wanted to keep his charitable works confined to his videos, he could've done just that. Instead, he's also set up events and even entire companies to handle his philanthropy.

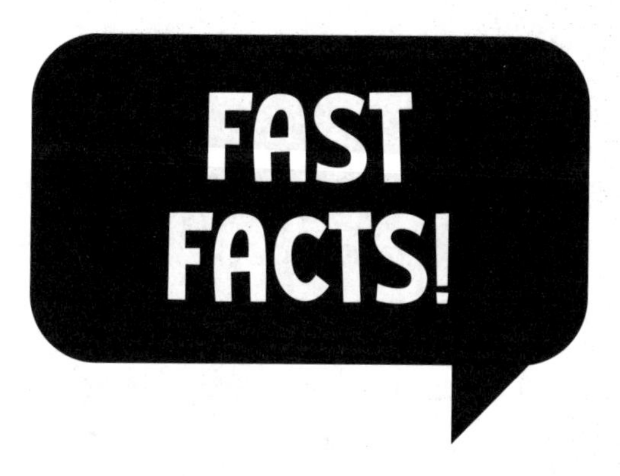

The word *philanthropy* means an act of goodwill toward the human race.

Philanthropy is also a term which is used to refer to a charitable organization.

Changing Lives

Although the cash giveaways are a good example of MrBeast's selflessness, he likes to think of new and different ways to express his generosity. That's where videos like "I Adopted EVERY Dog In A Dog Shelter" and "Giving $1,000,000 Of Food To People In Need" come in. They change lives in other ways (although the "Dog Shelter" video *does* also involve cash giveaways).

But if you've seen MrBeast's videos before, the "Food" and "Dog Shelter" videos might not seem that unusual. What he did in January 2023, however, was quite out of the ordinary in the best way possible. In a video uploaded that month and titled "1,000 Blind People See For The First Time," MrBeast paid for cataract surgery for a thousand patients. So while MrBeast may stage silly stunts involving giant bowls of cereal and underwater endurance tests, he's also someone who uses his platform to make the world a better place.

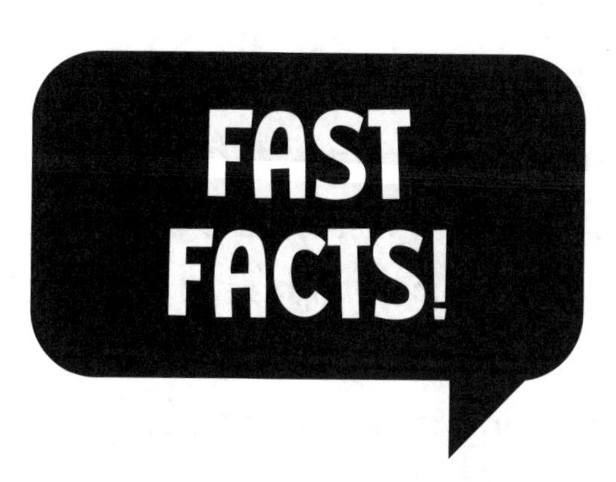

One man with low vision told MrBeast that he missed driving after his vision became blurry

So after that man's cataract surgery, MrBeast surprised him with a new car.

Team Trees

In October 2019, MrBeast launched a campaign called #TeamTrees. In order to both help the environment and celebrate reaching twenty million subscribers, he set the goal of raising $20 million, every penny of which would go toward planting trees. MrBeast hoped to make this happen by the end of 2019. Happily, the campaign met its goal even sooner than it had hoped, and it raised the full amount by Thursday, December 19, 2019.

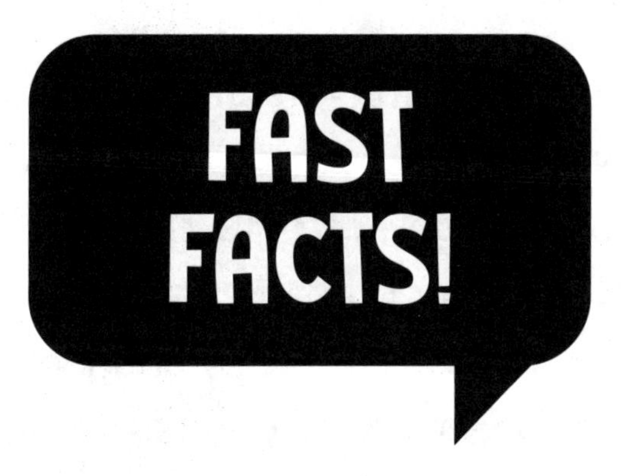

#TeamTrees was a partnership between MrBeast and the Arbor Day Foundation's reforestation program.

MrBeast kicked off the #TeamTrees campaign by donating $100,000 to it.

Team Seas

In 2021, following the #TeamTrees campaign, MrBeast started a YouTube channel exclusively dedicated to charitable efforts. This new channel was called Beast Philanthropy. As of the summer of 2023, the Beast Philanthropy channel had fifteen million subscribers. By the beginning of 2024, it had more than 23 million subscribers, showing that more and more people want to make this world a better place!

One of the efforts undertaken by the channel was the #TeamSeas campaign. MrBeast's goal was to remove thirty million pounds of trash from the ocean. He hoped to raise $30 million, and for every dollar raised, a pound of garbage would be eliminated. All of the proceeds went to two organizations: Ocean Conservancy and The Ocean Cleanup, both of whom clean up bodies of water. As of this writing, #TeamSeas has made sixty-three countries cleaner!

FAST FACTS!

MrBeast launched #TeamSeas with fellow YouTuber Mark Rober.

The #TeamSeas campaign ended up raising $33 million, exceeding its goal.

Did You Know That . . .

1. In the "1,000 Blind People See For The First Time" video, when some patients emerged from the eye surgery, he gave them suitcases of cash.

2. In one segment of the "I Donated $300,000 To People In Need" video, MrBeast pays the student loan debts for some college students.

3. In that same video, MrBeast buys MacBooks and PlayStation 5 consoles for a high school classroom.

4. Then he gives the teacher in charge of that classroom a backpack filled with $10,000 in cash.

5. Beast Philanthropy gives all of the money it earns to a warehouse, which provides mobile food donations all over eastern North Carolina.

6. According to *Business Insider*, by December of 2018, MrBeast had earned the title of "YouTube's biggest philanthropist."

7. MrBeast has also raised money for St. Jude Children's Hospital.

8. In his "Giving $1,000,000 Of Food To People In Need" video, MrBeast explained why it's so important for people to donate to food banks.

9. Many influential people donated to #TeamTrees.

10. They included former YouTube CEO Susan Wojcicki and MrBeast's fellow YouTuber PewDiePie.

Favorite Charity?

"I want to . . . use my main channel's influence to one day open hundreds of homeless shelters/food banks and give away all the money."

—MrBeast on his charitable work

What charity is nearest and dearest to your heart? Which one would you most like to support? Write about it on the lines below.

Helping Hand

Was there ever a time in your life when you needed help? Who was kind enough to give you that help? A family member? A friend? A classmate? Write about it on the lines below.

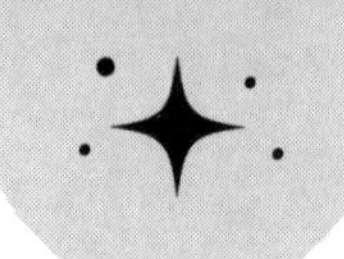

Quiz: Sympathy and Empathy

1) **MrBeast's "I Adopted EVERY Dog In A Dog Shelter" video took him nine ____ to film.**

   a. centuries
   b. millennia
   c. eons
   d. months

2) **The two organizations which partnered with MrBeast for #TeamSeas divided up their efforts to clean bodies of water. Ocean Conservancy cleans oceans and ____.**

   a. castles
   b. beaches
   c. fortresses
   d. palaces

3) **And The Ocean Cleanup cleans ____.**

   a. rivers
   b. snowballs
   c. snowmen
   d. snow globes

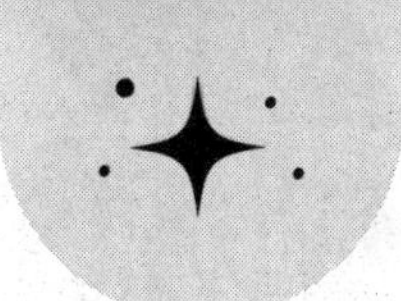

4) Aside from being a YouTuber, #TeamSeas cofounder Mark Rober is also a former ____ ____.

a. SHIELD agent
b. Monarch agent
c. NASA engineer
d. Time Agent

5) MrBeast's fellow YouTuber Jaiden ____ donated $20,000 to #TeamTrees.

a. Clock-watcher
b. Elevator operator
c. Lamplighter
d. Animations

Check your answers on page 78!

BEAST PLANET

Entertainment Empire

Today MrBeast works from a massive studio complex near his hometown of Greenville, North Carolina. He runs a sprawling multimedia empire consisting of eighteen YouTube channels, including the main MrBeast channel, as well as MrBeast 2, MrBeast Gaming, BeastReacts, Beast Philanthropy, and various international channels.

The videos on MrBeast's main channel are the more expensive ones, so that channel is financially supported by the Gaming and Reacts channels. The videos on those two channels prominently feature the Beast Gang and are less expensive, so they bring in a huge amount of money.

As of early 2024, MrBeast's main channel has more than 253 million subscribers, and his videos—in total—have gained over 30 billion views. What does that mean? It means that MrBeast makes fun, engaging, inventive content that's different enough to stand out from the pack. Viewers like his laid-back, everyman quality, as well as his sense of humor and his generosity toward those less fortunate.

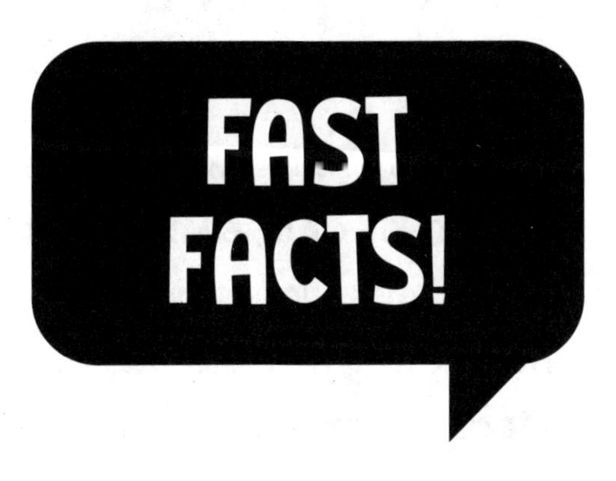

The studio complex where MrBeast currently works is worth $10 million.

It occupies one hundred acres of land and includes several warehouses where he shoots his videos.

Don't Forget to Like and Subscribe

Where will MrBeast go from here? He's currently in the process of answering that very question. That's because right now MrBeast's company is constructing three huge content and production centers in Greenville, North Carolina. Why?

Because his dream is to make Greenville a haven for content creators working in the digital media space. Will he realize this ambition? That remains to be seen.

But MrBeast went from a teenager whose videos didn't get very many views to an entrepreneur whose videos typically get millions of views (sometimes more) and who owns his own megasuccessful business. So one thing he's proven is that he has the drive and the talent to make his dreams a reality.

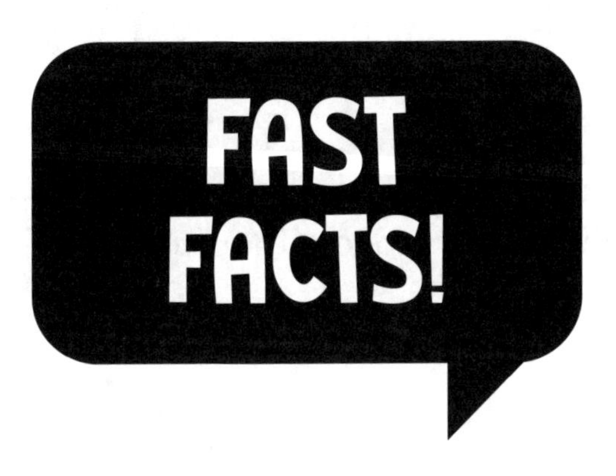

In both 2022 and 2023, MrBeast won the Nickelodeon Kids Choice Award for Favorite Male Creator.

When MrBeast won in 2022, in keeping with Nickelodeon tradition, a geyser of green slime hit him in the face.

Did You Know That . . .

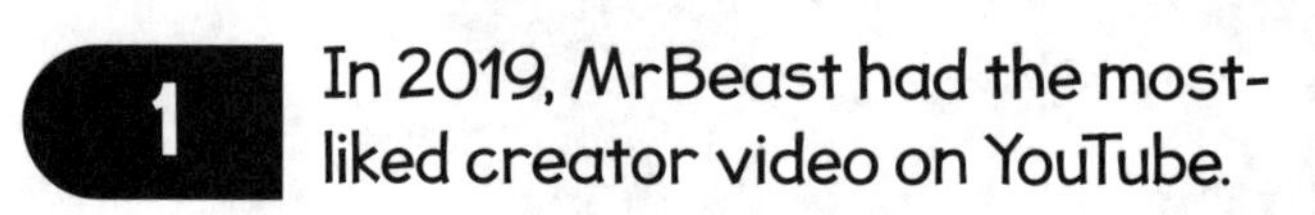

1. In 2019, MrBeast had the most-liked creator video on YouTube.

2. Fittingly that video was titled "Make This Video The Most Liked Video On Youtube."

3. By the end of 2020, MrBeast was the No. 1 most-subscribed creator on YouTube in the United States..

4. When MrBeast accepted his Nickelodeon Kids Choice Award in 2022 (the year he was slimed), he said that his mother has kept every award he's received.

5. He explained that this is something she's done ever since he was a child.

6 When MrBeast won his Nickelodeon Kids Choice Award in 2023, he descended from a UFO hovering above the stage.

7 In August 2023, MrBeast had a cameo in the animated film *Teenage Mutant Ninja Turtles: Mutant Mayhem.*

8 In the film, he played a character who is listed in the credits as Times Square Guy.

9 In January 2024, *Variety* reported that MrBeast was in talks to develop a reality-competition TV show for Amazon's Prime Video streaming service.

10 Not much is known about the show as of this writing, but its format would mirror MrBeast's YouTube challenges.

Always Ambitious

"This is the tip of the iceberg. Give me 20 years and then see what we will accomplish."

—MrBeast on his ambitions

What's *your* greatest ambition? What would you like to accomplish, either now or when you're an adult? Write about it on the lines below.

Your Own Business

If you had your own business, what kind of business would it be? What would you make? Would it be a restaurant? A movie studio? An art gallery? A comic book store? A sporting goods store? A detective agency? Or some other type of business? Write your answer on the lines below.

ANSWER KEY

Pages 18–19:
1) c, 2) a, 3) d, 4) b, 5) c

Pages 34–35:
1) b, 2) a, 3) c, 4) d, 5) b

Pages 50–51:
1) d, 2) a, 3) c, 4) b, 5) d

Pages 66–67:
1) d, 2) b, 3) a, 4) c, 5) d

ABOUT THE AUTHOR

Arie Kaplan wrote the six-volume Shockzone: Games and Gamers series of children's nonfiction books, which covered every aspect of the video game industry. He also wrote the award-winning nonfiction book *From Krakow to Krypton: Jews and Comic Books*, which was a finalist for the National Jewish Book Award.

Aside from his work as a nonfiction author, Arie has written numerous books and graphic novels for young readers, including *Jurassic Park Little Golden Book*, *Frankie and the Dragon*, *LEGO Star Wars: The Official Stormtrooper Training Manual*, *The New Kid from Planet Glorf*, *Batman: Harley at Bat*, *Spider-Man Comictivity*, *Shadow Guy and Gamma Gal: Heroes Unite*, and *Speed Racer: Chronicles of the Racer*. In addition, Arie is a screenwriter for television, video games, and transmedia. Please check out his website www.ariekaplan.com.